Rhyming Polemics Vol 2

JON KAUFMAN

Jon Kaufman is a former London based history teacher and trade union organiser who eventually gave up the classroom and the picket-line in order to turn his table tennis hobby into a full-time job. As a table tennis manager, he led his club to an unprecedented ten successive British Premier League titles, whilst simultaneously developing a wide ranging model of social inclusion in sport.

When not fully immersed in matters Ping, Kaufman turned his attention to the blogosphere, where his social and philosophical views could be given an airing. He continues to coach and promote table tennis across the country whilst working on publishing a trilogy of polemical material.

The first part of his Rhyming Polemics trilogy, *Alphabet Soup*, lays out Kaufman's general philosophical outlook, often in sharp polemical style. What is Alphabet Soup polemicising against? Petty prejudices, ancient hatreds and all forms of stultifying parochial tribalism. There is also a consistent broadside against the destructive aspects of global corporatism.

This, the second instalment, '*Rhymes From A Dying Planet*', focuses exclusively, as the title suggests, on the global ecological catastrophe now spectacularly unfolding in front of our very eyes. Cleverly employing different voices and viewpoints, the rhymes seek to lay bare our conflicted attitudes towards what may well become an existential crisis. The body count, argues Kaufman, is rising day by day.

CONTENTS

First published in Great Britain as a softback original in 2022

Editing, design, typesetting and publishing by UK Book Publishing

www.ukbookpublishing.com

ISBN: 978-1-915338-45-7

Cover Photo © Kerry Rawlinson on Unsplash

Smoke image © Viktor Talashuk on Unsplash

INTRODUCTION

Dear reader,

If you're looking for sublime poetry, you're looking in the wrong place. In fact, if you're looking for poetry of any real quality, I doubt this is the anthology for you. All I can offer you is a polemic, pure and simple, blunt and direct. Because, if we really are in an existential moment, a sharp polemic is most certainly what is required.

OK, I offer my polemic with a rhyme, but it's certainly not poetry in any artistic sense of the word. The rhyme has been added merely to make the thing a little more palatable. Because let's face it, a polemic is, in essence, a lecture, and nobody likes to be lectured. Not me, not you, and certainly not those of a youthful disposition, battling away at school, college, university or perhaps languishing somewhere on our demotivating low-wage/benefit system. And it is to this youthful constituency that this anthology is principally aimed.

Why so? Because young people tend to be less compromised, less trapped by the matrix of the modern world. Entrapment comes soon enough. But before debt, wage and consumer slavery take full hold, there is a brief window where one might just glimpse something else. Something more communal, less corrupting, less demeaning, less destructive.

The inspiration for this little collection comes principally from the renowned Nigerian writer, Ben Okri. In an opinion piece in The Guardian, 12/11/21, Okri argued that all writers, all artists, should focus their efforts on exposing the coming climate emergency. Write, he said, as if we really are in End Days. So before I hand you over to Mr Okri, who puts the case so eloquently, I conclude my opening remarks by wishing you, my dear readers, some amusement, some discomfort, and in the end, perhaps some motivation. I will be content with achieving any one of the above.

Artists must confront the climate crisis – we must write as if these are the last days – BEN OKRI

Full article published in The Guardian 12/11/21

Faced with the state of the world and the depth of denial, faced with the data that keeps falling on us, faced with the sense that we are on a ship heading towards an abyss while the party on board gets louder and louder, I have found it necessary to develop an attitude and a mode of writing that I refer to as existential creativity. This is the creativity at the end of time.

Albert Camus, writing during the second world war, felt the need for a new philosophy to answer the extreme truths of the times. The absurd was born from that. Existentialism was born too from a world in the throes of extreme crisis. But here we are on the edges of the biggest crisis that has ever faced us. We need a new philosophy for these times, for this near-terminal moment in the history of the human.

It is out of this I want to propose an existential creativity. How do I define it? It is the creativity wherein nothing should be wasted. As a writer, it means everything I write should be directed to the immediate end of drawing attention to the dire position we are in as a species. It means that the writing must have no frills. It should speak only truth. In it, the truth must be also beauty. It calls for the highest economy. It means that everything I do must have a singular purpose.

It also means that I must write now as if these are the last things I will write, that any of us will write. If you knew you were at the last days of the human story, what would you write? How would you write? What would your aesthetics be? Would you use more words than necessary? What form would poetry truly take? And what would happen to humour? Would we be able to laugh, with the sense of the last days on us?

We have to find a new art and a new psychology to penetrate the apathy and the denial that are preventing us making the changes that are inevitable if our world is to survive. We need a new art to waken people both to the enormity of what is looming and the fact that we can still do something about it.

NAOMI KLEIN:
This Changes Everything

......we have not done the things necessary to lower emissions because those things fundamentally conflict with deregulated capitalism, the reigning ideology for the entire period we have been struggling to find a way out of the crisis. We are stuck because the actions that would give us the best chance of averting catastrophe are extremely threatening to an elite minority that has a stranglehold over our economy, our political process, and most of our major media outlets. P18

.......our economic system and our planetary system are now at war. Or more accurately, our economy is at war with many forms of life on earth, including human life. P21

....underneath all of this is the real truth we have been avoiding: climate change isn't an "issue" to add to the list of things to worry about..... It is a civilizations wake-up call. A powerful message, spoken in the language of fires, floods, droughts, and extinctions, telling us that we need an entirely new economic model and a new way of sharing this planet. Telling us that we need to evolve. P25

Fear is a survival response. Fear makes us run, it makes us leap, it can make us act superhuman. But we need somewhere to run to. Without that, the fear is only paralysing. So the real trick, the only hope, is to allow the terror of an unliveable future to be balanced and soothed by the prospect of building something much better than many of us have previously dared hope. P28

"As far as I'm concerned, the struggle to end poverty and inequality and the struggle to address climate change can, must and should be seen as two sides of the same coin."

KUMI NAIDOO:
Human rights and environmental activist

RHYMES FROM A
DYING PLANET

THE REPORT

The IPCC just warned us that our planet is slowly dying, but the climate crisis deniers say the scientists are really lying, but when I switch on my TV and watch the forests frizzle frying, I want to curl into a tiny ball, it's all so utterly mortifying, and when I watch the glaciers melting there can be no more denying, and as I slump into my chair you can hear me sadly sighing, even though I try my best to be so ecologically complying, while the corporates pump out fakery and increase their falsifying, about the speed of global warming which just keeps on multiplying, now we're told the Gulf Stream may collapse, which is beyond all speechifying, although our governments talk a lot about electric green supplying, but it may be much too late is what I'm basically implying, because those coal-fired power stations is what they're insanely still applying, though the advance to clean renewables is really highly gratifying, and the rise of Extinction Rebellion is really very satisfying, and I'm re-reading Naomi Klein, which is clear and edifying; 'This Changes Everything' her message loudly amplifying, that action is urgently needed before the planet gets a hiding, so the conclusion is very simple, it's a scientific finding, that Wall Street is the culprit and it needs an occupying, nothing short of Occupation or our planet will be dying, and the IPCC report is clever in correctly identifying, all the dangers that we face and the problems underlying, this is science at its very best, not merely casual prophesying.

THE HEDGEHOG

Hello, I'm a hedgehog, and I've got myself
embroiled in a whole heap of trouble
There used to be a hedgerow here, but all I can
find is this whole heap of rubble
I loved that old hedgerow where I could slumber all day, and
in the night-time root around in my own little bubble
Though I was not much impressed with the fields all around,
not much to eat for a hedgehog in that insecticidal stubble

Anyway I digress from my heart-rending story of existential woe
It transpires they're building a bypass to greatly speed the flow
So my farm with its lovely hedgerows got sold off months ago
And now I'm feeling most vexatious, because at this
point in my story I have nowhere left to go.

Now I appreciate you're busy with important
things upon your mind
I don't seek to blame you personally because
you're generally quite kind
But with the decline of my habitat I'm in something of a bind
And where life used to be so rosy now my life is quite a grind.

But please don't imagine for a moment that I'm deeply in despair
I've a plan gestating nicely to create a brand new homely lair
But first to navigate this trunk road with excruciating care
Because the death of a single hedgehog is hardly something rare
Though the death of all my hedgehog friends
should make you most aware
That your human-centric activities are increasingly unfair
So before I bid goodbye to you, I offer you a dare
Reform your destructive ways right now so
we hedgehogs get our share.

Look I hope I haven't depressed you with
my gloom and whimsicality

Though I'm starting to conclude that we're
    near the point of criticality
And whilst I offer up this rhyme in genuine species solidarity
You may wish to ponder the morality of
    your shrinking hospitality
Because I am tired of all your platitudes
    and your blatant criminality
And your arrogance and your ignorance and
    your much hyped humanality
Now please help me across this busy road,
it's my immediate practicality.

NOT MY FAULT

Bish, bash wallop, what a commotion
Clever old humans in full throttle motion
It's not my fault that we killed off the ocean
And all other species got a short sharp demotion

Clean fusion energy and all the renewables
But coal, gas and oil are much more doable
It's not my fault that the air is unbreathable
The end of the world? Totally inconceivable

Push things to the limit and to hell with the consequences
If the seas get higher just build higher fences
It's not my fault if it ups the expenses
An ecological crisis? It's good for the senses.

Holy Mary, the fish are all dead
Better start eating plastic instead
It's not my fault if I ain't well read
Not interested in the science, I won't be led.

Black cat, white cat, just catch those mice
Love those insecticides, don't think twice
It's not my fault about the melting ice
Let's go for broke, just roll those dice.

Mother of God, the hills are on fire
Chuck on some rubbish and let the flames fly higher
The whole world's looking like a funeral pyre
It's not my fault if we go to the wire
Know another planet for sale or hire?

THE BICYCLE

If you want to start to change the world just jump upon a bicycle
And suddenly the whole wide world is
feeling a lot less hierarchical
And you'll cease to be the emitter of that polluting nasty particle
But don't be tempted by those battery bikes,
with their prices astronomical
Just stick to the original bicycle, they're
the genuine healthy article
Or try the latest fold-up type, they're most
convenient and remarkable

To cycle is simplicity, it's not overly mechanical
It's speedy and it's practical, and you'll be
considered quite the radical
And you'll save yourself your hard-earned cash
because they're really most affordable
So give two fingers to all those firms that
make breathing so intolerable
With their polluting predilections and
their profits most deplorable
Just jump upon your bike today and strike
a blow against Big Capital
Then we can all survive sustainably and live happily ever afterble.

EXTINCTION

Nothing, nothing at all. All gone. Quite devoid of presence. A total and complete absence of being. A condition of ceasing to exist. They were here but now they are not. Just like that. In the blink of a Homo sapien eye. Gone.

They were multi-coloured, just in case you were interested. Exotic. We're talking Amazon but not the shopping kind. They may have had cousins. No one is sure. Too late anyway. Either way, it's done and dusted. Kaput. End of the line. Yesterday's story. Ancient history. Get the picture? Of course you do.

They had their uses. Very useful as it goes. Food source for an equally exotic winged creature. An Amazonian flying thing. Oh well, plenty more insects in the forest, to coin a phrase. Mustn't complain. These things happen. Extinction is all the rage, I'm reliably informed. Can't get enough of it. Here today, gone tomorrow. What a lark. A walk in the park. A taste of the dark. Just leaving our mark. OK, we should have seen it coming. Should have built an ark. Still, what a lark.

UNCLE SAM

'The business of America is business', said President Coolidge in a voice loud and gruff, and we invade other countries where we cut it up rough, because we're after the minerals and all the valuable stuff, and we love that black gold of which we can't get enough, and our talk of democracy is just humbug and bluff, and we don't give a damn about your humanistic piffle and we don't give two hoots for your environmental puff, and if you dare to protest you'll end up in a cuff and if you persist in your endeavours you'll go under a bus, and you should know it by now that we play the game tough, and there's no time for morality, no time to discuss, because the bottom line must be worshipped without fretting or fuss, because it's all about profit which is A double plus, and Uncle Sam is the boss and you'll just have to adjust, and regarding the planet, in Uncle Sam you must trust.

WORKER ANTS

Eight billion worker ants in the human nest
Consume all in sight and to hell with the rest
Get busy being born then on with the quest
Live a full life and do so with zest
Don't fall by the wayside, that's part of the test.

Things to get done and comply without question
Then gulp down your food and to hell with digestion
Obey the consensus, make no smart suggestion
Don't hesitate or contemplate, dare no digression
Then apply all your talents with fierce animal aggression

But why are some worker ants more equal than others?
Should we regard the whole nest to be our sisters and brothers?
Apart from my work is there more to discover?
Why do the Queen ants take a series of lovers?
Should I take a big chance and sneak out from the covers?

Eight billion busy bees in the human hive
It's a long and tricky struggle just staying alive
Cooperation is the key if the hive is to thrive
But we'll need a higher collective consciousness
 before that happy day arrives.

BLAH

From tomorrow we'll plant a billion trees
And clear the plastic from the seas We'll
pass some brand new eco rules
We'll ban for good those dirty fuels

We love to listen to that fine young Greta
And reading Ms Klein couldn't get much better
Of course we still love listening to Big Al Gore
And we promise to chop the forests no more

We really do love our little blue planet
This precious piece of spinning granite
And we're planning to build back better these days
A Green Revolution is our latest craze.

We'll hold a conference to sort it out
We know what's what so have no doubt
Fear not young souls of flood or drought
Just trust your leaders, no need to shout
We'll give that climate change a rout
We really do know what it's all about.

TECHNOLOGY

Short of drinking water? No problems, I'll build
you a solar-powered desalination plant.
Don't worry about the spiralling costs, just apply
for a non-existent government grant.

Too much pollution in the atmosphere? Worry not,
I'll build you a giant extraction machine.
And before you can say Naomi Klein, your air
will be sparklingly pure and pristine.

Worried about all the junk in the oceans? No problem,
I'll invent an enzyme to eat up the plastic.
And before you can down a bottle of cola, all the
oceans will be renewed and fully fantastic.

Troubled by the decline in the planet's biodiversity? Fear
not, I'll clone all the endangered plants and creatures.
Just give me a sample, living or dead, and I'll replicate
them all with their original features.

And what about the long-term survival of we homo
sapiens. Are we going to make it, I hear you ask?

Don't worry your busy little heads about that, the
robots are coming, it's my very next task.

COP 26: THE BUILD UP

They're coming from across the world, from
many far-flung destinations
From nations big and small, and the global corporations
They're coming to save the planet, for us and future generations
They'll be Biden, Bo Jo and Bolsonaro, ready to
make their enlightened declarations
There might even be Modi, Putin and President Xi
to enhance our collective expectations
And from Down Under we'll hail Scott Morrison,
he'll surely add much to the celebrations
Then the new batch of German politicians will
undoubtedly make some worthy recommendations.

There's no need for any agitation, no need for demonstrations
No need to ruin the party vibes with any childish remonstrations
Just stay at home and watch TV whilst
they hold their deliberations
And if you believe in Greater Powers above,
you might try some incantations
But I'm confident the world is in safe hands,
and they'll meet their obligations
And the flooding and droughts and raging fires
are just nature's annoying provocations

The New Year will bring us all much joy
and a most sunny constellation
Of ecological global rules and some sparkling innovations
And I steadfastly refuse to criticise, or
hear any words of deprecation
About our heroic national leaders and
their ingenious calibrations
So hip hip hooray and three cheers to all,
for their many inspirations
May they bathe forever in the sunshine of
our grateful acclamations.

COP26: THE PROGNOSIS

Good Cop, Bad Cop, it's difficult to say, with capitalism still conducting itself in the same nefarious way, and the banks still funding fossil fuels forever and a day, and all those fancy political declarations which will shortly fade away, because our leading politicians are entangled in the oligarchal sway, whilst the rabbis, priests and mullahs can only procrastinate and pray, so we better get ourselves together and we better join the fray. No point in moaning over lunch, no point in nodding nay, when we all have culpability in this existential play, so when our turn to speak our part arrives we had better shout out yea; yea to our survival in a new sustainable way, and yea to all the activists who are out there every day and most of all a resounding affirmation of nature's sparkling cabaret, but as we push towards the precipice I suspect my words are just cliché.

ECO WARRIORS

Let's get something straight right from the start
An Eco Warrior is imbued with a razor-
sharp mind and a battle-hard heart
They're the best of the best, almost a species apart
And like the suffragettes before them they
are primed for their part
Taking aim at their targets with the flight of a dart
Campaigning with precision like a fine piece of art.

An Eco Warrior may hail from any location
They vary in age and in their economic station
They engage in battle without the least hesitation
Workers, peasants and intellectuals of every gradation
Risking their lives amidst great personal privation.

Eco Warriors are the advanced guard of a coming war
Waged by the elites against the marginalised poor
Standing in defiance against corrupt corporate law
They declare by their actions, now listen to them roar.

PRIORITIES

No commitment to phase out the deadly black stuff
But Ireland beat the All Blacks in rugby today
No new commitments for rainforest protection
But England thrashed the Aussies in a most glorious display
Big Oil set to keep on drilling for oil
But Megs and Harry were seen at a thrilling ballet
No halt to the on-going biodiversity crisis
But Abba are celebrating their fiftieth bouquet
No attempt to reign in the agro-conglomerates
But the Queen is unwell so we'd better all pray
No commitments to honour the Paris Accords
But Prince William enjoyed his day at croquet
No attempt to assist the low-lying nations
But a new restaurant was opened to attract the gourmet
No agreement to keep below two degrees warming
But you really should see their amazing buffet
No timetable to halt the decline of the oceans
But I'm enjoying some fish at my favourite cafe
No sense of urgency that the planet's on fire
But at least I'm booked into my favourite winter chalet.

FOUR LIMERICKS

There once was a buffooning Old Etonian
Whose politics were somewhat draconian
He appeared like a clown
As he moped about town
And his climate policies were most un- Newtonian.

There's a Brazilian politician called Bolsonaro
And the agro-industrial complex is his close companero
He claims to be president
But his policies are malevolent
And we wish he would disappear up his own Kilimanjaro.

There's a Russian KGB president called Putin
And if you opposed him he'll sure put the boot in
He loves to sell oil
Bringing climate to the boil
While his oligarchs strut the planet highfalutin.

There's a newly elected US president called Biden
Who says he doesn't want the wealth gap to widen
He claims to be green
But that remains to be seen
Because his fossil fuel subsidies he's still cunningly hidin'.

ICE

I am Ice and I shrink
I shrink in the face of heat
The heat of activity and
The human proclivity
For growth
Endless growth
Unsustainable growth
So I shrink
Terminally
Irrevocably
Existentially

I had a job
To be pristine
To be reflective
To be forever
But now I shrink
Not pristine
Not reflective
Not forever

An unvirtuous cycle
Human industrial heat
Shrink
Human consumerist heat
Shrink
Human technological heat
Shrink
Ad nauseum
I fracture
I splinter
Help
I reflect less
Help
I melt
Too hot
Shrink
Too late
Help
Shrink
Disintegrate
Die.

END DAYS

Can it really be the end of days, I rather think it not
Just another fleeting story that will quickly get forgot
The big boys will sort it out if things get really hot
No need for me to change my life, no need to lose the plot
I'll just relax and enjoy the ride, I like flying quite a lot
In fact I'd like it slightly warmer, that would really hit the spot
A touch of Mediterranean warmth yes
    please, I might even buy a yacht
Of course I'm not disputing the science
    geeks, I'm sure they know a lot
It's just I only have one little life, I've only got one shot
So if it's all the same to you and yours,
    I'll be keeping what I've got
I won't be marching in the streets, I won't join your big boycott
I'm going to live in the same old way, maybe win the state jackpot
And he who talks of end of days is just a crazy old crack-pot
You and I both know for sure, it's just a load of liberal rot
Can it really be the end of days, I rather think it not.

DENIAL

A sparkling new year full of hope and good cheer
Nothing to worry about, nothing to fear
The climate will stabilise, of that I am clear
Just keep marching forward, don't steer to the rear
Look out for yourself and those near and dear
Trust in the tech geeks with their brave new frontier
And all those dreary doomsayers will soon disappear

Sure, things have been tricky with some floods most severe
And the droughts and the wildfires have
been troublesome this year
But business as usual will soon reappear
And those boring climatologists need not interfere
Nor those do-goody activists who love only to sneer
Along with the experts who bleat in our ear
About ecological extinction so loomingly near

Well we refuse to be bamboozled by their statistical drear
It's time to rejoice in our best party gear
And march on triumphant to a fabulous New Year.

QUESTIONS

Mummy, Mummy will the elephants and
tigers really disappear for ever?
No darling, we will always be able to see
them in our lovely big zoo
Daddy, daddy, will it really be too hot to live on our planet?
No sweetie, it will always be just perfect for Mummy and for you
Mummy, why are the icebergs disappearing into the oceans?
Don't worry my honey, they'll be back in the
winter all sparkling and new
Daddy, a man on TV said all the insects are dying
Don't fret my little angel, I'm sure it's not true
Please teacher, why are there lots of people
protesting with placards?
Worry not children, its just a lot of silly
people creating a big hullabaloo
Greta, why do grownups keep talking about climate?
Ignore them young people, they're just making
plans that they'll never pursue
Dear Jesus, should we forgive the fossil fuel companies?
Definitely not my sweet children, they are run by bad
people who know precisely what they do.

AMNESIA

Our scientists keep warning we're drinking
in the last chance saloon
Yet still we engage in our mindless daily trivia

They tell us that human life could be extinguished like a busted
balloon From the Arctic to the burning deserts of Assyria

Yet in the face of all the facts we crawl
into our comforting cocoon
A case of humanity's evolutionary amnesia

But there is no plan B to live out on the moon
And no technology to be our everlasting saviour

Furthermore the avaricious corporations
that currently dominate the room
Are governed by motives that are most darkly ulterior

Meanwhile the media keep on bleating
out the same tiresome tune
All talk of extinction is just childish hysteria

And we ignore the hard science in favour
of our governing buffoons
As we chat happily in our local cafeteria

But we might usefully take cognizance of the coming typhoon
Least our legacy be nothing but slime and bacteria.

WASTE

Wear it once then throw it away
Burn it, dump it or send to Malay
This is our world of work, rest and play
Mountains of plastic piled higher each day

'Reuse and recycle', it's a catchy refrain
But the recycling industry is all about gain
It's cheaper to incinerate than reuse again
And reducing consumption goes against the capitalist grain

So who stands to gain from this consumerist blight
No prizes for guessing it's about corporate might
A handful of conglomerates too powerful to fight
And they export our waste junk plain out of sight

Now our rivers and oceans are clogged full of debris
But our consumerist lives still fill us with glee
Even our solar system is no longer detritus-free
So we all wait in vain for that ecological decree

I once read that Bangladesh was the ultimate recycling nation
An unexpected honour borne out of chronic privation
Every item that can be salvaged goes into recirculation
And for this super human effort they deserve a global citation.

ANTHROPOCENE

Do you find the science of climate change so really very boring
I'd rather read a good book by a wood burner brightly roaring
Who wants to study graphs with their indices forever soaring
When you could be snuggled up in bed with
your partner gently snoring
Or perhaps watch your team slugging it out, cheering
out loud for some last second scoring

As for me, I'm rather keen on some urban style exploring
And play some ping pong with my mates
which I still find so adoring
Anything but those scientists with their faces so imploring
About the anthropogenic age of human activity so deploring
If we chop the odd few trees, we can plant
a few more in the morning
Get a life and get out there, ignore the scientific outpouring
We humans are too busy for nature's tiresome restoring
All the rest, as I said, is just so tedious and boring.

BEAUTIFUL BEACHES

A few years back I holidayed down on the Med; golden
sands and warm waters were awaiting for me
There was only one caveat as far as I was concerned, an
oil slick had covered me from my head to my knee
But the local authorities were extremely considerate,
they provided scrubbing brushes entirely for free
So I tiptoed in back through the slime and the sludge
determined to enjoy my day at the sea

Then a few years later I was on the Adriatic coast,
with an island tour a must you'll agree
I hopped through the islands without a care in the world,
until confronted by mountains of burning debris
It was the start of the season and the locals were busy,
burning the plastic that had washed in from the sea
It was a local initiative to clean up their beaches
without waiting for any tiresome official decree

More recently I visited the beautiful Canaries;
to escape the winter is a real jamboree
To lie on the beach with some tapas in hand,
and an ice cold beer is a hot guarantee
But my holiday was interrupted by an army of sellers,
Their impoverished homelands they were desperate to flee
And on arrival they were forced into merciless work
gangs, selling their stuff with a pitiful plea
I had entered the world of fortress Europa,
growing fascistic by increasing degree
And with poverty exacerbated by flooding and drought,
the next chapter of this story is clear to foresee.

CARBON DIOXIDE

How very typical of our contrary old Universe, how
annoyingly and vexatiously so predictable
That there is enough carbon dioxide for the cycle of life,
but too much and things become most inhospitable
You can argue all day about the merits of growth, but
excessive CO2 makes the planet uninhabitable
The balance in our atmosphere must be maintained at all
costs, that is a fact that is indisputably incontrovertible
The solution is staring us right up in the sky, to
ignore it would be outrageously inexplicable
To dilly dally for even a nano second or more, might
be suicidal and scientifically inadmissible
But as a species we seem hell-bent on defying
the balance of nature, each passing day
renders the results more unpredictable
How amazing then that the Universe even offers us a
chance, to act rationally or be insanely undialectical

TREES

You really don't get it, you guys from the West
With your pure vegan diets which you think are the best
Drinking shakes made of soya as you embark on your quest
To see who's most holy, to see who's most blessed

But remember us peasants that harvest your crop
Working all day till we're plain fit to drop
We don't get a break, we don't dare even stop
In fear of all the landlords, in fear of the cops
Try giving it a go, perchance care to swap?

Yes, it is true we cut down the Amazonian trees
Destroying the forests despite your heartfelt decrees
Of course we need forests, in that we agree
But our families must first pay the landlord his fees
So we cut down the forest ignoring your pious-most pleas

So as you set out on your quest, can you give guarantees
That you'll change the world order and
    stop the corruption and sleaze
Can you reign in the corporates that
    plunder the land and the seas
Can you offer us a lifeline to get up from our knees

If the answer is affirmative we will cease cutting those trees
And change into guardians and forest trustees.

WAR AND PEACE

Another dirty little war, another catalogue of misery
Geopolitical red lines drawn amidst claims of breached security
Whilst the 'war on terror' rumbles on,
though who questions its validity
But some wars go unheeded if the oil is far too miserly
Or the victims are the Other with their
complexion lacking purity
Meanwhile the threat of annihilation points
to our stubborn immaturity

But there's another war that's raging,
another form of state barbarity
A war against the planet and its bio-rich diversity
Spurred by infantile consumption within a capitalist conspiracy
And the masters of that universe are embroiled
in greed and rank duplicity
Whilst those that toil both night and day
are doomed to dreary poverty

I would like to join a war, a war against kleptocracy
To march with banners flying high to defend our slim democracy
To take up arms against no man, but rage against hypocrisy
Then expropriate ill-gotten gains from the ruins of autocracy
And seek out equilibrium amidst the chaos of cosmology
And commune awhile with nature's gold
and its vibrant rich ecology
That is the war I wish to wage, to be in
the vanguard of its soldiery.

TOP DOG

Still sitting with my laptop doing the usual sort of pondering
With my few remaining brain cells doing
their usual sort of wandering
What shall I write about, what shall I create
My thoughts are all over, my energies still a-squandering

What can I say about the environment today
Should I stay in and write or go out to play
Should I take up the fight or just stay home and pray
What if we humans have really had our day

They tell us the dinosaurs once ruled across the lands
Roaming hither and thither on their own and in bands
They were kings of the castle from the mountains to the sands
Perhaps they get a bad press because we just don't understand

Now we clever old sapiens are considered top dog
We know how to build and we know how to blog
We're the product of evolution from the slippery slimy frog
But our time might be up as we choke upon our smog
So we now have to ask; was it really worth the slog.

SOLAR

I'm a star in every sense of the word
And I know what is fake and what's true
I disappear every evening like clockwork, and
in the morning re-emerge into view
I'm a friend of your planet, but don't get too close,
or like Icarus you'll melt into glue
But don't go too far from my loving
embrace, or I'll turn you depressingly blue

I never cease to be amazed by stupidity,
particularly the stupidity of man
I've been waiting most patiently year after year,
those fossilised fuels you're refusing to ban
I'm offering you a deal, it's the deal of the century,
to make it part of an overall plan
To power your planet with my super charged rays,
but your rulers just don't give a damn

I can desalinate your oceans with hardly a cost
and your citizens will have water to spare
And I can heat up your homes and cool them in
summer with everyone getting their share
I can energize your industries without creating a
mess and soon you'll have hardly a care
It's plain common sense, not much to debate,
you just need to become more aware

But don't wait too long to make up your minds,
you've only a dozen years left
My offer is honest and simplicity itself,
there isn't that much to digest
Just stop drilling for oil and digging up coal,
the big boys you'll have to arrest
I'm inviting you all to embrace my whole plan,
step up my esteemed Earthly guest.

OIL AND WAR

Oil and war fit together like a horse and carriage
They fuel each other like the perfect marriage
They march together, don't dare disparage

Those tanks need oil to keep them rolling
Those planes need oil to keep controlling
Pump that oil for the troops enrolling

The machines of war need constant oiling
The bombs need oil to do their spoiling
To unleash their destruction to keep things boiling

Military might needs lubrication
Pour in the oil to prevent constipation
There's money to be made from aggravation

The missiles are primed for detonation
We're all fixated on our own pure nation
Time perhaps to envisage a new creation.

WHAT DO I KNOW

Just what do I know about planetary warming
In all honesty not a diddly damn thing
And what do I know of the Anthropocene
Only that the word has a nice sounding ring
What can I say about the sixth great extinction
Well we homo sapiens are still surely king
And what are the chances to alert the whole world
Whilst we're all enjoying one happy last fling

What do I know about glacial retreat
Not very much I'm most certainly sure
And what do I know about the bleached coral reef
Only that the topic is almost certain to bore
And what can I say about all flooding and droughts
Well I doubt if there's any known cure
And what can I add to what is already well known
Not so much as a pile of manure

So what do I know about the capture of carbon
Well it's not a speciality of mine
And what do I know about deforestation
Please join me for a fine glass of wine
Just what is the sum of the things that I know
May the sun in the sky always shine
So in conclusion let me sleep with my head in the sand
Don't bother me I'm doing just fine.

INSULATE

What's your home like buddy, mine's a tad cold and draughty
If I could just get it fully insulated that
would be so cool and crafty

I've read that 30% of CO2 derives from heating up our home
But if we leave the heating off my friend,
we get chilled right to the bone
And when the bills come rolling through the
door, the whole country starts to moan
The government just shrugs its shoulders and
says, 'you're on your own Jack Jones'.

I'm seriously thinking of protesting, and
joining 'Insulate Britain'
They're blatantly breaking the laws each
day to get the laws rewritten
Their arguments are very persuasive, they've got me fully smitten
And they're all so wonderfully mischievous,
like a three-month playful kitten

Insulation may not be the ultimate solution
More like evolution than a full-blown revolution
But it would create a lot of jobs you know, and
make a mighty powerful contribution
To the transition to green renewables,
which is the obvious resolution.

CARS

Just who are you to tell me whether I should or shouldn't drive
My car gets me to work each day without which I can't survive
It's more than just a job you see, it's the way I stay alive
And if I wait for that old omnibus, it's bound to not arrive

Just who are you to tell me I had better use my bike
I won't be bossed about by you, I'll do precisely as I like
You seem to have an obsession, are you
some demented sort of psych
If you're so against the motorcar, I suggest you take a hike

Just who are you to tell me that I should convert to full electric
Your views are really jumbled up, you're quite the mad eclectic
If they see me driving hybrid I'll be considered most eccentric
Get yourself together friend, you're far too ego-centric.

Just who are you to tell me to give up my precious car
Without four wheels beneath me I won't be getting very far
Do you honestly see me walking to my favourite drinking bar,
It's time for getting serious pal, it's time to cut out all the blah
I love my little box on wheels, I love my motor car

END OF THE WORLD

Haven't humans always protested that
we're all doomed and set to die
They've pointed to deadly pandemics and
giant meteorites from the sky
Every generation has had its prophets pronouncing
Armageddon is nearly nigh
But in the morning we're eating breakfast
and in the evening some apple pie

But this time it could be different, I hear
our learned high priests say
We have bundles of scientific evidence, this time we'll have to pay
We must pay the price for ignorance, we have clearly lost our way
The planet's near the precipice, we are near our final day

But surely our scientists can save us, they're
really a very clever bunch
These men and women who devote their lives
to preventing the final crunch
They toil all day in their laboratories,
never daring to stop for lunch
But the tipping points are getting close, it's
hard science not just a hunch

So I've resigned myself to accept this fate that
we humans have had our time
The way we trashed our earthly home, it's a really heinous crime
It seems our best days are behind us now
Perhaps we're heading back to slime
But I won't go quietly without a fight, I'll
compose another rhyme.

JOURNEYS

You might say we're all on a journey
Though the route is rarely clear
Marching blindly from crisis to crisis
Feigning hope over inner fear
Is it a material or metaphysical journey
Either way it's hard to steer
Past economic insecurities
Rents forever in arrear
Just surviving is considered a victory
As our lives become austere

It might be a journey for national identity
More fraught with every year
Or a journey with religious overtones
Though we atheists tend to sneer
Or our journey may be gender orientated
Whether straight or somewhat queer
But there is no escaping the journey
Through our troubled earthly sphere

But our journeys have a collective dimension
Whose destination might be somewhat near
I refer to the journey for harmony
With nature's wondrous green frontier
A journey for all humanity
In which we have been so cavalier
Riding roughshod over fauna and flora
As eco-systems disappear
If we ignore this part of our journey
The consequence will be severe
Like the floods and fires in Australia
Or the shrinking forests of deep Zaire.

THINKING

I went merrily camping quite recently in
my gas guzzling campervan
I was after some well earned R&R, well
that was the generalized plan
In the mornings I explored the riverside
walks, that's how my day began
And in the afternoons I'd taken to taking a nap,
in order to recharge the inner man

Then I'd wake up thinking about the world,
wondering if we had a chance
To keep temperatures below those two degrees,
which is considered the logical stance
But with coal and oil still on the roll we
were doing a doomful dance
It seemed we were determined to burn our world,
or so it appeared at a casual glance

But my thoughts soon took a more positive
turn, reforestation was in full swing
And a switch to renewables was gathering pace,
clean air they were sure to bring
And the technology to capture the carbon
was a thoroughly inspiring thing
And what about those new electrical cars,
pretty soon they'd be rolling in

But again I started to get those negative vibes,
as grim reality started to bite
Like the flytipping along the riverside paths, which
was such an ugly incendiary blight
So too the network of motorway lanes which
defiantly refused to be quiet
And that industrial incinerator burning our waste,
belching emissions by day and by night

I realised then that to have even a chance we
would have to learn to collectively fight
To hold to account the criminal class, their
crimes we must surely indite .

LAST PERSON STANDING

Will the last person standing please turn out the lights
It's time to hand over to creatures less destructive
Your time in ascendancy has been relatively short
But your actions have proved most instructive
In fact you really have outstayed your welcome on Earth
To linger longer would be highly obstructive
I'm hearing your pleas that you really can change
And I admit your arguments are somewhat seductive
But we've weighed up the evidence concerning your case
And to allow you to continue would be counter-productive

Lights Out Please!

INTER-CONNECTED

I'm sitting here drinking my morning coffee in the sun,
wondering who picked my coffee beans and where
How much were they paid for their labours that day,
and was it exploitative and wholly unfair
And the apricot croissant that I'm enjoying so much,
what ingredients did they neglect to declare
Were there cancerous insecticides pumped into the
thing, would they tell us, I doubt that they dare
And what of the trainers I'm wearing today,
were their makers living a life of despair
Then I thought of all the things that I use back at
home, am I interested, do I really much care
I know everything is connected one way or another,
though I know I'm no clever Voltaire
I do know that pollution sweeps over the world,
and in the end we all breathe the same air.

CONTRADICTIONS

Today I'm planting a new sapling, and to be
honest I'm feeling pretty good
My tree is definitely not for chopping, not
destined to be turned into wood
It's part of world wide rewilding, to reforest
every barren neighbourhood
Why not take the plunge my ecological friend,
deep down you know that you should

We urgently need to plant some trees, plant
a billion by the end of the day
Time to rewild huge tracts of our landscapes,
other species could then have their say
We humans can live in smart cities,
sustainable in every which way
And let the hillsides and valleys bloom freely, a
place to walk and to rest and to play

But the motorway bypass they're building right
now, is important to speed up the flow
And the new Disney Land that they're constructing just over the
hill, promises to be a most spectacular child-centred show
And three cheers for the brand new shopping mall,
let the local economy prosper and grow
And I'm excited about our airport expansion, after
three lockdowns we're all ready to GO!

HOLLYWOOD

I watched a fun Hollywood film last week
A clever satire I'm sure you'll agree
A meteorite was hurtling towards Earth at great speed
But we all just carried on with our lifestyles carefree
Then I rewatched an old Hollywood favourite of mine,
A chilly tale I recommend you must see
It tells the story of global climate collapse
Where the temperatures drop by some considerable degree
But the Hollywood blockbuster that really hits home
Is a tale that just might come to be
Aliens have come down to save the future of Earth
'Not for us', but from people like you and like me
Keanu Reeves plays the alien with such magnificent control
He had me rooted to my comfy settee
All three films help to underline the dangers we face
Through satire and through sharp irony
Will they help to change attitudes and alter our course
I'm not hopeful for me or for thee.

DROUGHT

Mother and baby crying inconsolable tears
They say it's the worst drought for at least forty years

No milk in the breast so it's on with the crying
Very soon it'll be time for some drought-induced dying

The aid agencies are scarce, nowhere to be seen
The sight of dying children is acutely obscene

The crops are all dead, the animals too
Babies dropping like flies, what else can they do

A foreign TV camera crew creeps into sight
There's a story to be had from this East African blight

A cruel act of god or just nature's way
Nothing to be done at the end of the day

A politician is saying that the climate's to blame
More fossil fuel burning will bring more of the same

At the other end of the continent the floods still persist
The rains keep on coming, they refuse to desist

Meanwhile in India it's fifty degrees
No respite to be had, not even under the trees.

NOTHING TO SAY

What has Paul McCartney got to say
about the climate emergency
As it turns out absolutely nothing at all
What has Bob Dylan got to say about the melting of ice
I rang him last night but he refused to answer my call
What about Will Smith, has he anything to contribute
Not one word uttered in his American drawl
Surely Eminem has rapped about the state of the planet
But I can't find a statement either big or quite small
I bet Banksy has got something to say on the topic
But I couldn't find anything etched on the wall
Perhaps Her Royal Highness the Queen
has some pearls of wisdom
But nothing's coming up as I continue my trawl
I know, I'll try Elon Musk, the richest person of them all
But it seems it's not a priority as his empire continues to sprawl
In desperation I'll try ex-President Trump
But it doesn't figure in his twittering scrawl

So I'll just have to write me my own polemical rhyme
Maybe someone will read it way up in Nepal
And if I've done an injustice to any of the above
I'm happy to give you a call
To issue you an apology most sincere, be you
in London or sunny Senegal.

FOUR MORE LIMERICKS

There's a brave young woman called Greta
Who'd change the whole world if we'd let her
She knows what is what
And she won't take no rot
She's as determined as a galloping red setter

There's a dedicated young activist called Vanessa
From Uganda, they don't come much better
But she got chopped from the photo
From her head to her elbow
By the white media which has failed to impress her

There's a brilliant Canadian writer called Naomi
And she won't accept any greenwashing baloney
She's on to their tricks
As she hands out some licks
She's the real thing, she's definitely no phoney

There's a tenacious English journalist called Monbiot
Who's as radical as an enraged charging buffalo
He holds power to account
No mountain too high to surmount
From Westminster to far away Borneo.

VIRUSES

'Zoonotic spillovers' is the scientific term
It sounds pretty scary but you'd better soon learn
That viruses are coming to make humanity squirm
From species seeking escape from the planetary burn

We all thought that Covid was a nasty old bug
In fact it got to the point we couldn't give Granny a hug
But the next nasty virus could arrive from a slug
So in our post-Covid paradise I advise against being too smug

As the planet warms up some species are forced to regroup
They become desperate to survive so they'll jump through a hoop
To find a suitable space to eat and to poop
And they may inadvertently spread viruses into humanity's loop

A Zoonotic spillover is coming to a city near you
It may come from the wild or it may come from the zoo
Bats are the main culprits but any old critter will do
The next viral pandemic could be a lot worse than the flu.

OCEAN DEEP

What's happening in our oceans, I'm really not sure
But I feel fairly safe behind my tightly locked door
There's so much going on, what with pandemic and war
That I can't think about oceans, in fact I can't think anymore

But I'm told that acidification is a pretty big deal
Apparently absorbing CO2 makes for a rather indigestible meal
I'm told it's killing off species from the shark to the eel
Perhaps I'll ask a few fish just how do you feel

The warming of the seas is another point of contention
The science geeks tell us it takes things to a whole new dimension
These greenhouse gases seem like a real nasty invention
Yet we all seem intent on ignoring the Paris Convention

Matters are made worse, I've read, by plastic pollution
Dumping waste in the oceans is clearly not a viable solution
Keep this up and we're likely in for some real retribution
Maybe the polluters need punishment not soft absolution.

NARRATIVES

We all carry narratives one way or another
Perhaps it's 'love for thy neighbour' or 'death to the Other'
Of course, we inherit a narrative as we emerge from our mother
Though as we push on through, there's new stuff to discover

A narrative might tell us that all life is worth living
It might tell us that taking is inferior to giving
Alternatively it might teach us it's a dog eat dog world
Where there's no room for empathy, no room for forgiving

One narrative tells us we should look after our planet
After all, it's home turf, not just a cold piece of granite
But too often we abuse it and opt to say damn it
We've all got stuff to get done whether we're Ahmed or Janet

This destructive narrative tells us that we humans are boss
Destined to rule no matter the cruelty or cost
That all other species should carry the cross
Can we ever be rid of the slain albatross?

CREATIVE WRITING

I'm continually wrestling with fresh ways to argue the case
Need to step up to the task, it could be a life and death race
It's not as if we can migrate to a new home out in space
And if we get all this wrong we might disappear without trace

Ben Okri has pleaded that we should write
as if we're in 'End of Days'
Can't you feel the power of those skin blistering rays
Okri argues persuasively this is no mere passing phase
But like every other writer I feel trapped in a maze

How to hammer home the message without delivering a lecture
How to cease coming across as a mindless objector
How to summon the skill to create a grim haunting spectre
Of a doom-ladened planet like a clever film director

Most of my friends and associates vaguely agree
Like Newton, they get the analogy of the apple and the tree
They accept the science is as obvious as an A, B and C
But to focus exclusively on climate; "no that's
definitely not appropriate for me"

So a few of us plod on a little each day
Trying to say the same thing in a less tedious way
I keep banging my drum saying the polluters must pay
And I'm refusing to relent until my feet turn to clay.

HEATWAVES

Climate change has arrived with a cataclysmic bang
Across the entire Indian subcontinent they're feeling its fang
Despite all the dire warnings and attempts to harangue
The temperatures keep rising across poor Pakistan

The daytime temperatures are approaching fifty degrees
And the land surface temperatures are too hot for the bees
The local inhabitants are begging us please
Take this thing seriously, we're down on our knees

It's killing our crops and it's killing our land
The heat is unbearable, we cannot withstand
But our pleas go unheard, the response is so bland
Our bodies are boiling, we're turning to sand

This is no heatwave that arrives once in a while
It's becoming regular as clockwork and its effects are most vile
It sweeps over the subcontinent with its inhospitable smile
This is the future and it's killing in style.

THE WRONG RESPONSE

War cuts off gas so what do we do
Ring up Big Oil who assemble a crew
Start drilling for oil, profits anew
The nation is grateful, we'll all pull on through

But this response is so muddled to say the very least
We're getting it wrong from the West to the East
This is surely the time to say goodbye to the beast
Put an end to Big Oil, put an end to their feast.

Time to wean ourselves off the dirty black stuff
Let the corporates sulk, let them go off in a huff
Force them to change, call out their bluff
And if they refuse to play ball we can also play rough

Every crisis, they say, creates a new opportunity
We're an evolved clever species, time for human ingenuity
Bury the profit motive, build a global community
Big Oil is a killer with arrogant impunity.

AU REVOIR

This is the most uncomfortable question of all
Whether you're down in the Bronx or up in Nepal
Have we homo sapiens lost track of the ball
Do we deserve to survive; that's a difficult call

Let's face it, we've trashed the place from equator to pole
Consumed all in sight, that's been our primary goal
Fauna and flora both burned up like coal
Rapacious human beings rampaging out of control

So if you were visiting Earth from a planet afar
If you were looking down on Earth from a far away star
Would you regard human actions as somewhat bizarre
And just pack up your bags and wave au revoir

Would you be shocked at the things that we mindlessly do
Would you be surprised that our wealth
was in the hands of the few
Would you plead with our leaders to start things anew
Or just shake your head and bid us adieu.

M.A.D.

I half learned about it when I was a mere slip of a lad
Not entirely briefed but I knew it was bad
Something about war and we were all going mad
Dad said don't think about it, it'll make you feel sad

As I grew older I grasped Mutually Assured Destruction
All life would end, an end to all human construction
A nuclear winter would be the prevailing obstruction
And lethal radiation would be humanity's
only form of production

But slowly and surely we stopped fretting about M.A.D.
We moved on down the line to the next human fad
And like everybody else I was really quite glad
Until up popped a geezer from Russia called Vlad

So now we are lucky, we have two ways to die
Catastrophic climate collapse or atomic bombs from the sky
But we also have choices, to accept or defy
I think I'll opt for the latter before my final goodbye.

CHILD'S PLAY

It's not so difficult to compose your own rhyme
Just turn off the distractions and forget about time
Think about a stroll amidst the oaks and the pine
Blank out the stress, the city and the grime
And remind yourself that poetry is hardly a crime

It's not about cash, not a quid or a dime
More about the creation of something sublime
Reaching out for that feeling way back in your prime
When you had a sense of pure innocence
    without a mountain to climb
And before you know it you'll have constructed your rhyme
As easy as apples and lemons and lime

You may wish to attempt another stanza or two
Start to explore, give things a good chew
Stir it around like a mean witches' brew
Play with the language, let it ensue
Kick it about like the man in Kung Fu
Change colour midstream from violet to blue
And hey presto young poet, you'll have your debut

They'll be critics aplenty, no doubt about that
They'll call you a peanut, they'll call you a prat
They'll tell you to stick it right under your hat
They'll knock you right down, they'll knock you out flat
But you'll know deep down that they're a low stinking rat
Because you've made something original
    and they can't take that back.